E. Vinicius Camargo Barbosa
Geraldo Papa
Washington de Melo

Effect of different soil cover colours on pest control

E. Vinicius Camargo Barbosa
Geraldo Papa
Washington de Melo

Effect of different soil cover colours on pest control

Plastic Mulching

ScienciaScripts

Imprint

Cover image: www.ingimage.com

This book is a translation from the original published under ISBN 978-613-9-72176-4.

Publisher:
Sciencia Scripts
is a trademark of
Dodo Books Indian Ocean Ltd. and OmniScriptum S.R.L publishing group

120 High Road, East Finchley, London, N2 9ED, United Kingdom
Str. Armeneasca 28/1, office 1, Chisinau MD-2012, Republic of Moldova, Europe
Printed at: see last page
ISBN: 978-620-6-12400-9

I dedicate to my family. I love you.

THANKS

First of all, to God,
For health, wisdom and strength to continue my journey.

To my parents, Vanderlei and Susete and my sisters, Cristiane, Mariana and Luciana, for their unconditional love, supporting and advising me in every decision, always making me believe in my potential and never give up. We love you.

To my girlfriend and best friend, Lais, for her patience, complicity and constant dedication in my life. I love you.

To the supervisor, Geraldo Papa,
For the encouragement, patience and trust. I am grateful for the professional and personal advice, and also for making me believe in my ability.

To the Entomology II Laboratory Team, for the field knowledge transmitted to me, to my class and the others with whom I lived. Thanks for the materials and moments of relaxation.

To EMBRAPA INSTRUMENTAQAO, to Dr. Washington de Melo and his team. Thank you for the opportunity to improve my studies and move forward.

To the Universidade Estadual Paulista "Julio de Mesquita Filho". For the excellence in teaching quality, to the teachers, to Prof. Dr. Hdlio Ricardo and employees.

To the Rep. Ita Comigo "family", friends Heitor (Cachago), Vitor (Botelho), Ricardo (Pistao), Renato (Broxa), Gabriel (GG) and the dog (Pipa), true friendship prevails.

To my family, friends and other people who, directly or indirectly, have accompanied my trajectory and contributed to my maturing.

"The history of life on Earth has been a history of interaction between living things and their environments."

(Rachel Carson)

SUMMARY

The tomato thrips, *Frankliniella schultzei* (Trybom), is among the main pests of tomato cultivation, causing serious economic losses. The damage is due to the thrips feeding, causing lesions and transmitting the tomato *spotted wilt* virus (TSWV), known as "tomato head turner", causing a high cost in the control and developing resistance to insecticides, due to the need of sequential spraying. The present work aimed to evaluate the influence of ultraviolet rays reflected from different types of *mulching* materials used in tomato crop on thrips, under field conditions. The experimental design was randomized blocks, with eight treatments and four repetitions. Each plot consisted of a planted area of 3.2 m^2 , transplanted with AP 529 hybrid. The treatments consisted of the placement of polyethylene plastic in the colours: silver, black, white, red, yellow and gray, as well as rice husk and witness without soil cover. Before transplanting, the seedlings were treated with Mancozeb fungicide and no insecticides were applied to control tomato pests in order to evaluate the effect of ultraviolet rays on the thrips population. The evaluations were carried out at 14, 21, 28, 35, 42, 48 and 56 days after transplanting the seedlings, counting the number of thrips found in 4 random plants in each plot. The temperature and reflectance of solar radiation on the roof were also measured. At 56 days after transplanting, the number of plants with symptoms of virus infection and the number of fruits were counted. The data obtained were submitted to variance analysis by the F test and the means were compared by Tukey (5%). The treatment with a silver colour coating was efficient in repelling the adult thrips, reducing its population and the incidence of the TSWV "tomato head turn" virus in tomato crops and can be considered as another option for pest management.

KEY WORDS: *Lycopersicon esculentum,* ultraviolet, integrated pest management.

SUMMARY

CHAPTER 1

1. INTRODUCTION

The tomato crop *Lycopersicon esculentum* Mill. is one of the most important in the world, both for small and medium-sized farmers, for commercial and nutritional purposes. In Brazil, the planted area is approximately 65.07 thousand ha, with annual production of 4,291,160 million tons and average productivity of 65.9 t ha^{-1} , making it an important income generating product for Brazil (AGRIANUAL, 2016). Currently, Brazil is the largest producer of tomatoes for industrial processing and the largest consumer of tomato products in South America (EMBRAPA HORTALIQAS, 2006). In the last 20 years, Brazil has experienced promising results with tomato breeding programmes. The country evolved from a productivity of 37 t/ha in 1990 to 60 t/ha in 2010 (Brazilian Institute of Geography and Statistics - IBGE, 2012).

The tomato thrips, *Frankliniella schultzei* (Trybom), is among the important pests of the crop. Although there are approximately 5000 thrips species, only about 87 of them are considered pests (MOUND 1996). Thrips are the most efficient vectors in the transmission of *tomato spotted wilt* virus (TSWV) (SHERWOOD et al., 2001a, b). The symptoms of the virus are chlorotic and necrotic rings on young leaves, which later acquire a tan colouration and distortions in the lamina. When the disease reaches greater severity, the leaves and stem show necrotic areas and the death of the tip, causing a large reduction in plant growth and productivity. The curling of the leaves upwards is characteristic of the disease and responsible for the name "head turning". When the fruit is ripe it remains with chlorotic or necrotic ring spots, with pale red colouration, yellowish spots and concentric rings (COLARICCIO, et al., 2001).

The most important method of thrips control in tomato is the chemical one, using insecticides in sequential applications with an average of two sprays per week, which may reach up to fifteen sprays per crop cycle. However, thrips resistance to insecticides has been reported in several countries for most of the traditional classes of broad spectrum agrochemicals such as pyrethroids, carbamates and organophosphates (FUNDERBURK et al., 2011). The most efficient insecticides for thrips control in tomato are in the spinosyn class, but there are already reports documenting thrips resistance to spinosyns (WEISS et al., 2009).

Protected tomato cultivation aims to increase the productivity and quality of the fruits and can be implemented in small areas (MEDEIROS et al., 2009). The productivity is related to the leaf area index, which interferes with the interception of solar radiation by the plant canopy (PAPADOPOULOS; PARARAJASINGHAM, 1997). In this respect, when the soil is covered with plastic, it can constitute an important tool, because it can alter the balance of solar radiation for the plants and for the soil, depending on the optical characteristics of the material used (LIAKATAS et al., 1986, DECOTEAU et al., 1990, HAM et al., 1993, HATT et al., 1994). The aluminized cover that reflects ultraviolet rays is very effective in reducing the colonization of thrips species of the genus *Frankliniella* and the incidence of tomato leaf curl (MOMOL et al., 2004). The objective of this work was to evaluate the influence of ultraviolet rays reflected by different *mulching* materials on the tomato crop on thrips under field conditions.

CHAPTER 2

2. BIBLIOGRAPHIC REVIEW

2.1 THE TOMATO

It is speculated that the tomato started being used as food in the USA around 1850, the first known cultivar being the ponderosa.

In Brazil, the tomato was probably introduced by Italian and Portuguese immigrants, and its cultivation started in the 1940s (EMPRESA DE PESQUISA AGROPECUARIA E EXTENSAO RURAL DE SANTA CATARINA - EPAGRI, 1997).

Tomato processing for industrial purposes started in Southern Italy and the United States more than a century ago. However, in the last 30 years this activity has experienced an improvement in agricultural and industrial production, particularly in the 1990s (BRANDAO; LOPES, 2001).

Tomato is of great economic importance for the country, and stands out for its nutritional value, characterized by its high content of carotene, thiamine, niacin and vitamin C (17-24 mg/100g of fresh fruit). The fruit is rich in lycopene, the substance responsible for the red colouring, recommended for the prevention of prostate cancer (BOUER, 1999).

Tomato is a dicot, order Tubiflorae, belonging to the Solanaceae family (FILGUEIRA, 2003), genus *Lycopersicon* (CAMARGO, 1992; SILVA; GIORDANO, 2000), subgenus *Eulycopersicum f*MINAMI; HAAG, 1989), species *Lycopersicon esculentum* (PAZINATO; GALHARDO, 1997). The name Mill came from MILLER who, in 1754, was the first to propose the botanical classification and the name *Lycopersicon* (MINAMI; HAAG,

1989).

The genus *Lycopersicon* encompasses nine species, and can be grouped into two complexes according to whether they can cross *(esculentum* complex) or not *(peruvianum* complex) with *Lycopersicon esculentum.*

The *esculentum* complex comprises seven species: *Lycopersicon esculentum* Mill, Lycopersicon *cheesmanii* Riley, Lycopersicon *pimpinellifoliun* Miller, Lycopersicon *chmielewskii* Rick, Lycopersicon *parviflorum* Rick, Lycopersicon *hirsutum* and *Lycopersicon pennellii* (SILVA; GIORDANO, 2000; ARAGAO et al., 2002). The majority of the species are wild and not exploited, because their fruits are extremely small, often pubescent, but they are used in tomato breeding programs for the introduction of genes, which confer resistance to pests and diseases, improving the quality of the fruit and its nutrition (GARDE; GARDE, [1993?]; LOURENQAO et al, 1997; SILVA ; GIORDANO, 2000; ZORZOLI; PRATTA; PICARDI, 2000; ARAGAO et al., 2002).

The species *Lycopersicon esculentum* has a fleshy and juicy red berry when it reaches physiological maturity, with two (bilocular) or more lobes (plurilocular), which may reach twelve lobes (MINAMI ; HAAG, 1989; CAMARGO, 1992; GARDE; GARDE, [1993?]; FONTES ; SILVA, 2002). The ovary wall is called pericarp where it has three layers: exocarp, mesocarp and endocarp. The exocarp and endocarp are epidermal tissues, the mesocarp is not epidermal, because it involves the placental tissue where the seeds are, which are fertilized ovules (SOURCES; SILVA, 2002).

2.2 TABLE TOMATOES

L. esculentum is one of the most cultivated in the world, because it presents different varieties within the species to meet the diverse demands of the table tomato market and for industrial processing (SILVA; GIORDANO, 2000; FONTES; SILVA, 2002).

After the potato, tomato is the second most important vegetable in terms of economic importance, but the first in terms of popularity. While the potato is consumed in larger quantities only in the south and southeast regions, tomato occupies a prominent place almost everywhere in the country (SONNENBERG; SILVA 2004).

Among the cultivars for the market, it is desirable to choose those with resistance to diseases, such as *Verticilium, Fusarium* and nematodes, because when they are present they determine the failure of the crop. The size and uniformity of the fruits should also be taken into account (Fontes and Silva, 2002).

Among the table tomato varieties, they are classified according to the shape of the fruit into two groups: oblong, when the longitudinal diameter is larger than the transversal one, and round, when the longitudinal diameter is smaller or equal to the transversal one (BRASIL, 1995; BRASIL, 2002), corresponding to the commercial cultivars, Santa Cruz and salada (kaki or apple).

For tomatoes intended for processing, production is done with prices previously agreed in contracts between producers and industrialists, while for table tomatoes the market is free, with strong seasonality in prices and quantities. The main distribution channels in Brazil are the regulated warehouses and retail chains (CAMARGO et al., 2006).

The production of table tomatoes in Brazil has undergone major technological transformations in the last decade. One of the main adjustments is the segmentation of cultivars in the field (GUALBERTO; BRAZ; BANZATTO, 2002). The tomato cultivars that are produced for commercialization are characterized by their shape, number of lobes, coloration, type of growth of the aerial part (determinate and indeterminate) and resistance to packing and transport (SOURCES; SILVA, 2002).

Consuming tomato fruits contributes to a healthy, well-balanced diet. They are rich in minerals, vitamins, essential amino acids, sugars and dietary fibres. Tomatoes contain large amounts of vitamins B and C, iron and phosphorus. Because it is a crop with a relatively short cycle and high yields, tomato cultivation has good economic prospects and the cultivated area is increasing every day (NAIKA et al., 2009)

2.3 TRIPES - *Frankliniella schultzei* (TRYBOM, 1925), Thysanoptera: Thripidae 2.3.1 DESCRIPTION AND BIOLOGY

Thrips are small insects, which makes them difficult to identify. They can reach a maximum length of 1 mm and a wingspan of 2 mm. The nymphs, after hatching, are light coloured and measure 1 mm in length.

Adults may be darkly coloured from yellow to brown (ALVARENGA, 2004). Its wings are relatively long and fringed. The legs are lighter than the body and abdomen and have 10 segments, ending with a curved ovipositor in females. Oviposition may occur on the tenderest parts of the plant and each female oviposits from 20 to 100 eggs. Hatching of the nymphs occurs at 4 days and this phase can last from 5 to 10 days, depending on temperature.

Before becoming adults, thrips remain immobile for 24 hours on the plant or soil. They live inside the flowers, ventral side of new and old leaves, and shoots.

They form colonies and feed on sap, and the eggs are laid on the parts of the plant where they later attack. Thrips is a polyphagous species and can easily transfer from crop remains or weeds to the tomato plant.

These infestations may occur up to 45 days after transplanting or establishment of the new plantation, and the longevity of the insect is 20 days on average. The proliferation of this pest is favoured in hot and dry periods, but it can also appear in low temperatures associated with drought (ALVARENGA, 2004).

2.3.2 DAMAGE

Thrips are scraper-sucking insects that become infected by the tomato attack virus (TSWV) by feeding on the sap of diseased plants and inoculating the pathogen into healthy ones.

The thrips causes the greatest impact when the attack occurs in seedlings, but the symptoms are already apparent after transplanting, causing losses to the producer right from the start of the crop's establishment.

The losses are considerable and, depending on the infestation and the time of year, may compromise the entire production (GALLO et al., 2002). In the adult stage, if the insect is contaminated by the virus, when carried by the wind to seedlings in seedbeds and tomato crops already planted, they suck the healthy plants and inoculate the disease (ZUCCHI, 1993). Thrips should be sampled in the upper third of the plants, by tapping the tops of five plants per sample point, on a white tray, in a total of 20 points per plot. The level of control is one adult insect/plant (ALVARENGA, 2004).

2.4 TOSPOVIRUS

Frankliniella spp. species are the most efficient vectors of *tomato spotted wilt virus* (TSWV), this virus is one of the 20 species known as Tospovirus (SHERWOOD et al. 2001a, b). The most characteristic symptoms of Tospovirus are chlorotic and necrotic rings on young leaves, which acquire a distorted tan colour. In a more advanced phase of the disease, the leaves and stem show necrotic areas and death of the tip, besides a drastic reduction in growth with a reduction in productivity.

The characteristic curling of the leaves is responsible for the name given to the symptoms caused by the virus on tomato plants.

The mature fruits are left with chlorotic or necrotic ring spots, with pale red colours and yellow spots, besides concentric rings (COLARICCIO, et al., 2001). From the agricultural point of view, *Frankliniella shultzei* (Trybom) and the thrips *Frankliniella occidentalis* (Pergande) are important for causing damage to stems, leaves, flowers and fruit of tomato, and indirectly transmitting the tomato head turner, causing a drastic reduction in productivity. Tomato crops have been attacked by the disease, with more than 50% of the production compromised (STRECK, 1994).

2.5 INTEGRATED THRIPS MANAGEMENT

Thrips control should be preventive, giving priority to the seedling production phase. The seedlings should be produced in protected locations (green house) and/or with chemical control of this insect-vector, when an average of one adult insect is found per stem. Normally, the control of this insect in sowing is done with systemic insecticides or by applying insecticides from the neonicotinoid group, such as imidacloprid (Confidor 700 GRDA - 200 $g.ha^{-1}$, with 10 to 15 mL of syrup applied to the plant as a spray), or the neonicotinoid acetamiprid (Mospilan - 250 $g.ha^{-1}$, with 1000 L of $syrup.ha^{-1}$) (GALLO et al, 2002).

Most tomatoes that are treated with neonicotinoid insecticides at planting with soil applications for the control of *Bemisia tabaci,* whitefly, such as Imidacloprid, Thiamethoxam or Dinotefuran that are commonly used, can suppress thrips foliar feeding, but will have limited impact on thrips in flowers. Broad spectrum insecticides, such as Pyrethroids, which are used for control of whitefly adults can reduce natural enemies of thrips. Pymetrozine is effective against whitefly adults and does not interfere with thrips management. Lighter molecules, such as Azadirachtin-based products and microbial insecticides such as Beauvaria bassiana, and growth regulators to control whitefly, are compatible with thrips management. (RILEY; PAPPU, 2000, 2004, RILEY, 2008; KENNEDY, 2008; OLSON, 2008; RILEY et al, 2009a;. RILEY et al, 2009b).

In Spain the spinosin chemical group is widely used for the control of lepidoptera and thrips, after being introduced in 2002, with an initial optimum control of *Frankliniella occidentalis* (ESPINOSA et al., 2005), in some cases excessive use, with at least ten applications per crop, has produced highly resistant populations in some fencerows in southeastern Spain (BIELZA et al., 2007). In addition, subsequent applications of broad spectrum insecticides help suppress the important natural enemies of other pests, which effectively suppress mites, whitefly, miners and other pests (FUNDERBURK et al. 2000, REITZ et al. 2003, SRIVISTAVA et al. 2008).

2.6 PLASTIC COVER

Polyethylene mulches have been used since the 1950s to increase the yield of horticultural crops (EMMER, 1957). Advantages of plastic mulching include earlier harvest as higher yield per unit area, improved fruit quality, more efficient use of nutrients and water, reduced matocompetition, and a potential decrease in pest insects and phytopathogens (LAMONT, 1993). The influence of mulches on plant microclimate and energy balance is a function of transmittance, absorbance, and reflectance of solar radiation (HAM et al., 1993; LAMONT 2005; TARARA 2000).

Ground cover started to be used worldwide on a large scale with the appearance of polyethylene plastic films, because of their low cost and ease of application, as well as other advantages they bring to crops. In Brazil, it has been used for many years in strawberry cultivation (GOTO, 1997).

2.6.1 METALIZED PLASTIC

Thrips colonization behavior can be disrupted by incorporating a mulch that reflects ultraviolet light, thereby reducing thrips numbers on host plants (BROWN; BROWN, 1992; KIRK, 1997; KRING; SCHUSTER, 1992; SCOTT et al., 1989; STAVISKY et al., 2002). The use of a highly ultraviolet-reflective metalized cover as *mulching* provides the solar radiation intensity to disrupt the initial flight of thrips in the field (BROWN; BROWN, 1992; KRING; SCHUSTER, 1992; SCOTT et al., 1989). Once the plants grow, their foliage covers the *mulching,* cover the crop inter-row, so that reflectance no longer provides flight disruption (BROWN and BROWN, 1992; KIRK 1997).

New technology is being tested with the use of roofing materials that reflect ultraviolet (UV) radiation, and has been effective in reducing thrips

and TSWV incidence, as well as reducing whitefly, vertically integrating integrated pest management (IPM) in tomato. Management is complex and can involve a high cost in production. Metallised covering that reflects ultraviolet radiation *(UV- mulch)* can significantly reduce TSWV, but can delay fruit maturity, so can potentially affect the market price of tomatoes. On the other hand, resistant tomato lines can eliminate TSWV damage. Although there are various management and options such as reflective (UV) coverage coupled with chemicals, it is found that a combination of all these management alternatives provides better results in tomato cultivation (RILEY;
PAPPU, 2000, 2004, RILEY, 2008; KENNEDY, 2008; OLSON, 2008; RILEY et al, 2009a; RILEY et al, 2009b).

2.7 ELECTROMAGNETIC RADIATION

Electromagnetic waves propagate in a vacuum with the speed of light given by c = 3x108 m/s. Knowing the frequency of an electromagnetic wave (f) in a vacuum, the wavelength (!) of this radiation can be determined using the equation: . c,.

X = - (nm)

Light is a visible radiation that can be defined as being an electromagnetic radiation, it is capable of producing a visual sensation and that is comprised in a range of wavelengths (X) limited between 380 and 780 nanometers (GARCIA JUNIOR, 1996). Given the spectrum of visible light in this range of wavelengths can take on various colours, from violet to red.

2.7.1 ELECTROMAGNETIC SPECTRUM

Electromagnetic radiation is commonly classified by the frequency of the wave characteristic of each energy. The classification of electromagnetic radiation according to its physical effect leads to the construction of an electromagnetic spectrum ranging from gamma rays to radio waves (CAVALCANTE et.al., 2001). Among these, ultraviolet radiation (UV) stands out because it has important applications in the scientific field.

The discovery of this radiation occurred during the observation of the darkening of silver salts when exposed to sunlight. In 1801, the German scientist Johan Ritter noticed that the sun rays, just after the upper limit of the visible spectrum, i.e. invisible rays were able to oxidize silver halides. These were called deoxidizing rays (to highlight their chemical reactivity) and subsequently called ultraviolet light in the late nineteenth century (BALL, 2007).

2.7.2 ULTRAVIOLET RADIATION (UV)

The band of the electromagnetic spectrum corresponding to UV radiation (Figure 1) covers wavelengths from 1 to 400 nm and can be subdivided into vacuum ultraviolet - UVV (1 - 200 nm), far ultraviolet (200 - 300 nm) and near ultraviolet (300 - 400 nm). In general, photons in these wavelength ranges have different effects on matter. Another usual sub-classification divides UV radiation into UV-C radiation (100 - 280

nm), UV-B (280 - 315 nm) and UV-A (315 - 400 nm). UV-A radiation is the main type of UV radiation that reaches the biosphere from solar emission. The largest fraction of the other components of the UV spectrum is reflected or absorbed by the ozone layer, present in the stratosphere (SECURITY OF WORK; INPE, 2013).

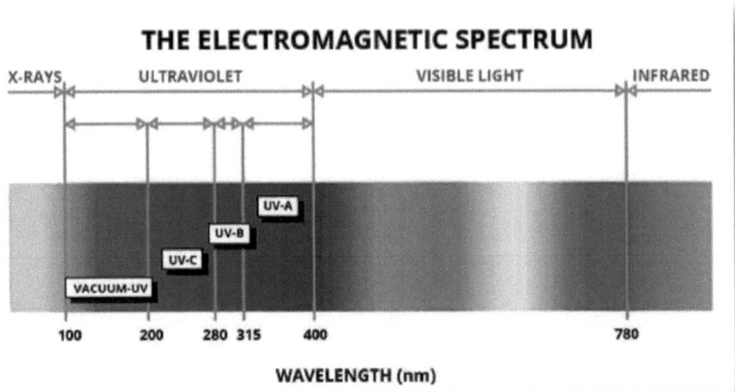

Figura 1. Electromagnetic spectrum, relating wavelengths to the respective classifications of electromagnetic radiation, with emphasis on the sub-classifications of UV radiation.

(Available at:http://www.semisyn.com/wp-content/uploads /2015/07/Ultraviolet- LEDs-Electromagnetic-Spectrum.png)

CHAPTER 3

MATERIAL AND METHODS

The experiment was carried out at the Fazenda de Ensino, Pesquisa e Extensao da Faculdade de Engenharia, UNESP, campus de Ilha Solteira, located in the municipality of Ilha Solteira - SP, in the period between August and October 2013. The geographical positioning of the trial site is 20°25'34.0" south latitude, 51°21'28.8" west longitude, with precision of 7m.The experimental area consisted of 5 lines of tomato plants, cultivar AP529, at spacing of 1.5 X 0.8 m. The characteristics of the soil at the test site is Red Distressed Latosol with pH 5.3 and 1.4% organic matter, the fertilization of planting was performed according to soil analysis.The statistical design was the randomized block design with eight treatments and four repetitions. The treatments considered were a control (no soil cover), rice straw *mulching* and polyethylene plastic mulching in silver, black, white, yellow, red and grey colours (Figure 2). The experimental plot consisted of one row of the crop with nine plants, considering the seven central plants as the useful area, totaling 3.2 m^2 each plot.The culture was conducted according to the recommendations for staked tomato culture, and then two applications of fertigation were made.The seedlings were treated with Mancozeb and there was no application of insecticide so that there was no interference in the evaluations on thrips repellency caused by the reflection of ultraviolet rays.

Figura 2. General view of the experiment, with eight types of ground cover for thrips repellency in tomato crop. Ilha Solteira-SP, 2013.
source: prepared by the author.

The evaluations were performed at 14, 21, 28, 35, 42, 48 and 56 days after transplanting, counting the number of thrips in each plot using white high density polyethylene trays with dimensions of 7.7 X 36.0 X 44.0 cm (Height X Width X Length). In each plot four plants were sampled at random, beating the plant on the tray five times.

At 35 days after transplanting, solar reflectance was measured at three random points in each plot at 30 cm from the ground using a Lutron Digital Luxmeter LM-8000. Also, the temperature of the *mulching* surface was measured at three random points using a Digital Infrared Thermometer with Laser Sighting (-50° to 380°C), (Figure 3).

Figure 3: Measurement of the *mulching* surface temperature.
Source: Photo prepared by the author.

At 56 days after transplanting, the number of plants with symptoms of virus disease in each plot and the total number of fruits produced in each plot were counted.

The ultraviolet reflectance of polyethylene plastics of the colours, white, black, silver, gray, yellow and red, besides rice straw and wet soil, was evaluated in the unit of EMBRAPA INSTRUMENTAL AO, located in the municipality of Sao Carlos-SP, using a spectroradiometer FildSpec3-ASD®, with detection capacity of 350nm to 2500nm spectral range (Figure 4, 5, 6 and 7).

The data obtained were submitted to variance analysis using the F test and the means were compared using the Tukey test at 5% probability. The original data were transformed into root of (X + 0.5) and the efficiency percentages were calculated by Abbott's formula (1925). The statistical package used was Assistat® (SILVA; AZEVEDO, 2002).

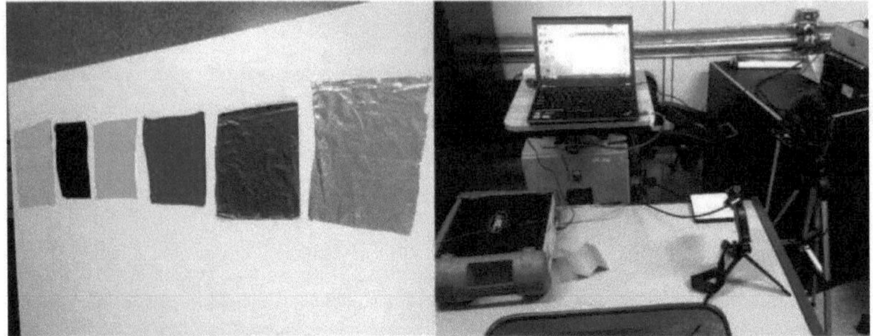

Figure 4. polyethylene plastics.
Figure 5: FildSpec 3-ASD radiometer spectrum.
Source: Photo prepared by the author.

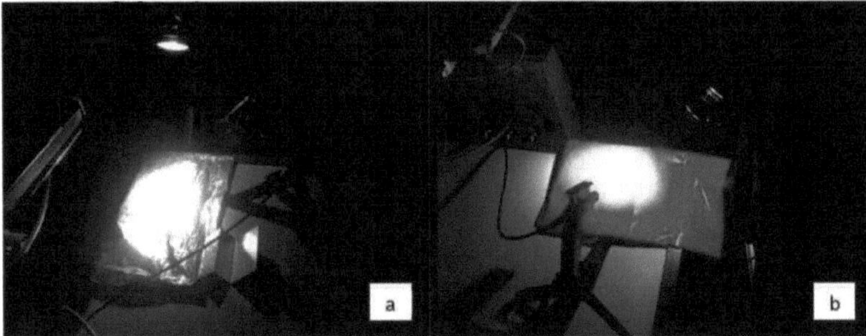

Figure 6: Time of evaluation on silver (a) and red (b) polyethylene plastics.
Source: Photo prepared by the author.

CHAPTER 4

RESULTS AND DISCUSSION

The results concerning the counting of the number of thrips in each treatment, carried out at 14, 21 and 28 days after transplanting are shown in Table 1 and the counts carried out at 35, 42, 48 and 56 days after transplanting are shown in Table 2. The treatment with silver plastic *mulching* provided pest control efficiency of more than 80% at 14, 21 and 28 days after transplanting. The treatment with rice husk did not differ from the control in any of the evaluations. The red and gray treatments did not differ from the black treatments in all evaluations, obtaining similar efficiencies. The white and yellow treatments did not differ from the control after 14 days. *Mulching* with white polyethylene plastic showed the highest incidence of thrips, being an attractive colour for the colonization of the pest. At 42 days, when the tomato plant leaves covered the *mulching,* because it was not staked, all treatments had low efficiencies, proving the action of ultraviolet reflected on the tomato leaves causing repellency of the adult thrips. The silver *mulching* treatment was the one with the best results in thrips repellency, because at 56 days after transplanting, the accumulated number of thrips was the lowest in all treatments, thus being the one that most reflects the ultraviolet rays to the tomato plants (Figure 7).

Table 1. Total number of thrips in each evaluation at 14, 21 and 28 days after transplanting.

Percentages of efficiency were calculated by Abbott's formula (1925). Ilha Solteira, 2013.

Treatments	14 Total	E%	21 Total	E%	28 Total	E%
1. Silver	2 b	91	13 d	91	38 c	82
2. Black	3 b	86	101 bc	27	198 b	6
3. White	44 a	0	453 a	0	687 a	0
4. Yellow	12 ab	45	235 ab	0	283 b	0
5. Red	7 ab	68	30 cd	78	117 bc	44
6. Rice husk	11 ab	50	234 ab	0	294 b	0
7. Grey	1 b	95	53 bc	62	153 bc	28
8. Witness	22 ab		139 ab		211 b	
CV% =	**91.01**		**56.78**		**30.07**	

[1] Means followed by the same letter in the column do not differ, by Tukey test, at 5% probability

Table 2. Total number of thrips in each evaluation at 35, 42, 48 and 56 days after transplanting.

Percentages of efficiency were calculated by Abbott's formula (1925). Ilha Solteira, 2013.

Treatments	35		42		48		56	
	Total	**E%**	**Total**	**E%**	**Total**	**E%**	**Total**	**E%**
1. Silver	67 b[2]	61	106 b	20	196 a	14	80 a	0
2. Black	69 b	60	121 b	9	234 a	0	102 a	0
3. White	465 a	0	391 a	0	270 a	0	101a	0
4. Yellow	176 b	0	160 b	0	276 a	0	98 a	0
5. Red	98 b	43	119 b	11	208 a	8	122 a	0

6. Rice husk	147 b	14	196 ab	0	234 a	0	92 a	0
7. Grey	66 b	61	127 b	5	246 a	0	72 a	0
8. Witness	171 b		133 b		226 a		62 a	
CV% =	**34.55**		**34.01**		**16.51**		**25.15**	

- Means followed by the same letter in the column do not differ, by Tukey test, at 5% probability.

In a study by Momol et al. (2004), using metal *mulching* that reflects ultraviolet rays, the authors concluded that there was a significant reduction in the incidence of the disease caused by the tomato head turn virus with or without treatment with insecticides.

In similar work, Funderburk et al. (2011) worked with mulch that reflects ultraviolet rays and concluded that it was effective in reducing thrips and whitefly population and the damage caused by these pests.

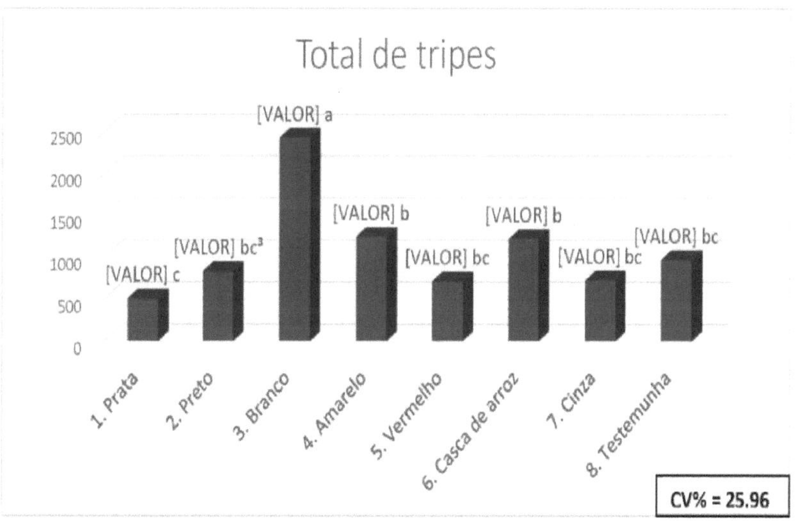

Figura 7. Accumulated number of thrips at 56 days after transplanting. The efficiency percentages were calculated using Abbott's formula (1925). Ilha Solteira, 2013.

[3] Means followed by the same letter in the column do not differ, by Tukey

test, at 5% probability.

In the evaluation of reflectance and temperature at 35 days after transplanting, the silver, white and yellow treatments had the highest indices, exceeding the scale of 2000 lux. The treatment with silver polyethylene plastic mulching was the one that obtained the lowest surface temperature, differing from the other treatments. The black plastic mulching obtained the highest surface temperature. The mean temperatures and light intensity are described in Figure 8.

When the total number of fruits per plot was counted, the treatment with silver plastic *mulching* obtained the greatest number of fruits (Table 3). In the control plants, thrips attack was intense, reaching viruses (TSWV) in 83% of the plants (Table 4). Due to the intense attack of thrips in the first month of evaluation, many tomato plants of the control died, therefore the number of fruits and plants with viruses was lower than in the other treatments.

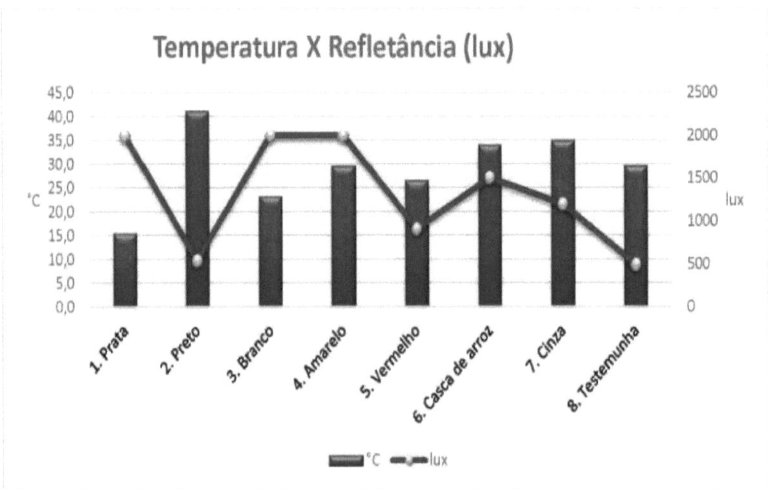

Figura 8. Evaluation of reflectance and temperature at 35 days after transplanting.

Table 3. Total number of fruits in each plot, at 56 days after transplanting.

Ilha Solteira/SP, 2013.

Treatments	Quantity of fruits per plot				
	A	B	C	D	TOTAL
1. Silver	342	26	298	210	1118 a[4]
2. Black	177	23	259	190	856 ab
3. White	11	14	301	101	562 ab
4. Yellow	94	15	239	395	886 ab
5. Red	268	19	134	336	935 ab
6. Rice husk	105	18	170	249	706 ab
7. Grey	247	22	182	287	936 ab
8. Witness	30	12	88	112	351 b
CV% =					**21.90**

[4]Means followed by the same letter in the column do not differ, by Tukey test, at 0.5% probability.

Table 4. Total number of plants with viraemia (TSWV) in each plot, at 56 days after transplanting.

Ilha Solteira/SP, 2013.

Treatments	Number of plants with symptoms of virus infection per plot				
	A	B	C	D	TOTAL
1. Silver	*2*	2	0	1	5 c[5]
2. Black	*2*	1	4	2	9 bc
3. White	*9*	5	9	7	30 a
4. Yellow	*3*	4	2	3	12 bc
5. Red	*1*	1	3	3	8 bc
6. Rice husk	*7*	3	4	3	17 ab
7. Grey	*2*	1	1	3	7 bc
8. Witness	4	3	3	3	13 abc
CV% =				**21.70**	

[5]Means followed by the same letter in the column do not differ, by Tukey test, at 0.5% probability.

The evaluations in the FildSpec 3-ASD® radiometer spectrum demonstrated that in the 350nm to 400nm spectral range, which corresponds
to the ultraviolet (UV) spectrum, the silver-colored polyethylene plastic was the one that most reflected the light intensity, followed by the gray-colored polyethylene plastic and the moist soil. The yellow, red and rice peel polyethylene plastics followed the same wave pattern, not possessing high light reflectance. The polyethylene plastic in the white colour increases its reflectance after the 360nm range, however it does not have the same reflectance intensity of the polyethylene plastic in the silver colour. The black polyethylene plastic absorbs light in the 355nm range, and has an uneven reflectance along the spectrum (Appendices 1, 2, 3, 4, 5, 6, 7 and 8).

CHAPTER 5

CONCLUSIONS

It is concluded that:

1- *Mulching* with white polyethylene plastic showed the highest incidence of thrips, which is an attractive colour for the colonisation of the pest.
2- Covering the soil with black, red and grey polyethylene plastic neither attracted nor repelled the thrips.
3- The covering of the soil with silver-coloured polyethylene plastic is efficient in reducing the initial colonisation of the thrips, providing a lower incidence of the pest on the plants and a lower incidence of plants with the "tomato head turn" virus disease.
4- Using the staked tomato production system, the reflectance effect of ultraviolet radiation on the tomato plant can be prolonged.
5- The use of silver polyethylene plastic *mulching* can be a new option for thrips management in tomato crops.

CHAPTER 6

REFERENCES

AGRIANU AL 2016. **Anuario da Agricultura Brasileira**. Sao Paulo: Informa Economics FNP, 2013.456p. 2016

ALVARENGA, M. A. R. Tomate: produgao em campo, em casa de vegetagao e hidroponia. Lavras: [s.n.],. p. 393. 2004.

ARAGAO, F.A.S.; RIBEIRO, C.S.C.; CASALI, V.W.D.; GIORDANO, L.B. Cultivation of tomato embryos *in vitro* aiming at introgression of *Lycopersicon peruvianum* genes into *L. esculentum.* **Horticultura Brasileira**, Brasilia, v. 20. n. 4. p. 605-610, 2002.

BALL, D. W. **Spectroscopy.** *3,* 14p. 2007

BIELZA, P.; QUINTO, V.; CONTRERAS, J.; TORNE M.; MARTIN, A.; ESPINOSA. P. J. Resitance to spinosad in the western flower thrips, Frankliniella occidentalis (Pergande), in greenhouses of south-aestern Spain. **Pests Management Science**, Sussex, v. 63, p. 682-687, 2007b

BOUER, J. Tomato is a good weapon against cancer. **Folha de Sao Paulo**, Sao Paulo, 2 May 1999.

BRANDAO, A.S.; LOPES, M.R. Cadeia do tomate industrial no Brasil. In: VIEIRA, R.C.M.; TEIXEIRA FILHO, A.R.; OLIVEIRA, A.J.; LOPES, M.R. ***Cadeias produtivas no Brasil:*** **Analise da competitividade,** Brasilia. Embrapa. Fundagao Getulio Vargas. 468 p. 2001.

BRAZIL. Ministry of Agriculture, Supply and Agrarian Reform. Ordinance No. 553 of 30 August 1995. Dispoe sobre a Norma de Identidade, Qualidade, Acondicionamento e Embalagem do Tomate in natura, para fins de comercializagao e Revoga as especificagoes de Identidade, Qualidade, Acondicionamento e Embalagem do Tomate, estabelecidas pela Portaria n°.

76, de 25 de fevereiro de 1975. **Diario Oficial da Republica Federativa do Brasil**, Brasilia, Sept. 1995.

BRAZIL. Ministry of Agriculture, Livestock and Supply. SARC/ANVISA/INMETRO joint normative instruction No. 09 of 12 November 2002. Provides on the regulation of packaging, handling and marketing of fresh vegetables, in packages suitable for marketing, for their protection, conservation and integrity. **Diario Oficial da Republica Federativa do Brasil**, Brasilia, nov. 2002b.

BROWN, S.L., BROWN, J.E. Effect of plastic mulch color and insecticide on thrips population and damage to tomato. **HortTechnology**, v. 2, p. 208-210. 1992

CAMARGO, L.S. **As hortaligas e seu cultivo**. 3.ed. Campinas: Fundagao Cargill, 253 p. (Sdrie Tdcnica, n. 6). 1992.

CAMARGO, A. M. M. P. de. et al. **Development of the tomato agroindustrial system**. Informagoes Economicas, Sao Paulo, v. 36, n. 6, p. 53-65, jun. 2006.

CAVALCANTE, M. A.; TAVOLARO, C. R. C.; **Caderno Catarinense de Ensino de Ffsica**, *18,* 298p. 2001

COLARICCIO, A; SOUZA-DIAS, J.A.C.; CHAGAS,C.M.; SAWAZAKI, H.E.; CHAVES,A.L.R.;
EIRAS, M. Novo outbreak of geminivirus em *Lycopersicon esculentum na* regiao de Campinas, SP.
Summa Phytopathologica. 27(1): 105, 2001.

DECOTEAU, D.R., KASPERBAUER, M.J., HUNT, P.G. Bell pepper plant development over mulches of diverse colors.**HortScience**, Alexandria v. 25, n. 4, p. 460-2, 1990.

EMBRAPA - BRAZILIAN COMPANY OF AGRICULTURAL

RESEARCH. A cultura do tomateiro (para a mesa). Brasilia: **Embrapa** - SPI, 92p. 1993.

Embrapa Hortaligas (2006). **Production systems. Cultivo de tomate industrial**. Available at: <http:cnptia.embrapa.br>. Accessed on 17 May 2016.

EMMERT, E.M. Black polyethylene for mulching vegetables. **Proceedings. American Society for Horticultural Science**. 69:464-469. 1957

COMPANY FOR AGRICULTURAL RESEARCH AND RURAL EXTENSION OF SANTA CATARINA - EPAGRI. Normas tdcnicas para o tomateiro tutored na regiao do Alto Vale do Rio do Peixe. Florianopolis: **Epagri,** (Sdrie Sistemas de Produgao). 1997.

ESPINOSA, P. J.; CONTRERAS, J.; QUINTO, V.; GRAVALOS, C.; FERNADEZ, E.; BIELZA, P. Metabolic Mechanisms of insecticide resistance in the western flower trips, *Frankliniella occidentalis* (Pergande). **Pest Management Science** v.61, p 1009-1015, 2005.

FILGUEIRA, F.A.R. **Novo manual de olericultura**: agrotecnologia moderna na produgao e comercializagao de hortaligas. 2.ed. Vigosa: UFV, 412p. 2003.

FONTES, P.C.R.; SILVA, D.J.H. **Produgao de tomate de mesa**. Vigosa: Aprendafacil, 197 p. 2002.

FRANTZ, G.; H. C. MELLINGER. Shifts in western flower thrips, *Frankliniella occidentalis* (Thysanoptera: Thripidae), population abundance and crop damage. **Florida Entomologist** v. 92, p. 29-34. 2009.

FUNDERBURK JE; REITZ S; OLSON S.; STANSLY P; SMITH H; MCAVOY, G DEMIROZER O; SNODGRASS C; PARET M; LEPPLA N. 2011. **Managing thrips and tospoviruses in tomato**. Document ENY-859 (IN895) Florida Cooperative Extension Service, IFAS, University of Florida.

EDIS. Available at http://ipm.ifas.ufl.edu. Accessed April 12, 2014.

FUNDERBURK, J., J. STAVISKY, AND S. M. OLSON. Predation of *Frankliniella occidentalis* (Thysanoptera: Thripidae) in field peppers by *Orius insidiosus* (Hemiptera: Anthocoridae).
Environmental Entomology v. 29, p. 376-382. 2000.

GALLO, D.; NAKANO, O.; SILVEIRA NETO, S.; CARVALHO, R.P.L.; BAPTISTA, G.C.; BERTI FILHO, E.; PARRA, J.R.P.; ZUCCHI, R.A.; ALVES, S.B.; VENDRAMIN, J.D.;

MARCHINI, L.C.; LOPES, J.R.S.; OMOTO, C. **Entomologia Agricola**. Piracicaba: FEALQ, 920p. 2002.

GARCIA JUNIOR, ERVALDO. **Luminotecnica.** 2ª ed. Sao Paulo: Editora Erica Ltda, 2002.

GARDE, A.; GARDE, N. **Culturas hortfcolas**. 6. ed. Lisboa: Classica, 469p. (Colegao Nova Colegao Tdcnica Agraria). [1993?].

GOTO, R. Plasticultura nos tropicos: uma avaliagao tdcnico-economica. **Horticiltira Brasileira**, v. 15, p. 163-165, Lecture. Supplement, 1997.

GUALBERTO, R.; BRAZ, L. T.; BANZATTO, D. A. Produtividade, adaptabilidade e estabilidade fenotipica de cultivares de tomateiro sob diferentes condigoes de ambiente. **Pesqiisa Agropeciaria Brasileira**, Brasilia, v. 37, n. 1, p. 81-88, 2002.

HAM, J.M., KLUITENBERG, G.J., LAMONT, W.J. Optical properties of plastic mulches affects the field temperature regime. **Journal of the American Society of Horticultural Science**. Ashford. v. 118, n. 6, p. 188-93, 1993.

HATT, H.A. et al. Influence of spectral qualities and resulting soil temperatures of mulches films on bell pepper growth and production. **Plasticulture**. Paris. v5. n.101, p. 13-22, 1994.

HTTP://satelite.cptec.inpe.br/uv/#/imgSig.jsp, accessed May 2016.

HTTP://www.segurancaetrabalho.com.br/download/rad-uv-seelig.pdf, accessed May 2016.

IBGE (2012). **Brazilian Institute of Geography and Statistics**. Disponivelem<http://www.sidra.ibge.gov.br/home/estatistica/indicadores/agropecuaria/lspa/ls p.Accessed on 10 May 2016.

KIRK, W.D.J. Distribution, abundance, and population dynamics, In: Lewis, T. (ed.). Thrips as crop P. **CAB International**, Wallingford, UK. p. 217-257, 1997.

KRING, J.B., SCHUSTER, D.J. Management of insects on pepper and tomato with UV- reflective mulches. **Florida Entomologist**. v.85, p. 9-14, 1992.

LAMONT, W.J. Plastic mulches for the production of vegetable crops. **HortTechnology** 3:35-39. 1993.

LAMONT, W.J. Plastics: Modifying the microclimate for the production of vegetable crops. **HortTechnology** v. 15, p. 477-481, 2005.

LIAKATAS, A.; CLARK, J A.; MONTEITH, J.L. Measurements of the balance and soil heat flux.
Agricultural and Forest Meteorology. Amsterdam. v.36, n.5, p.227-239, 1986.

LOURENQAO, A.L.; NAGAI, H.; SIQUIERA, W.J.; USBERTI FILHO, J.A.; MELO, A.M.T. Selegao de tomateiros resistentes a toposvirus. **Bragantia**, Campinas, v. 56, n. 1, p. 21-31, 1997.

MARTINS, M. B. G. ; CASTRO, P. R. C. Aspectos morfoanatomicos de frutos de tomateiro cultivar Angela Gigante, submetidos a tratamento com reguladores vegetais. **Bragantia**, Campinas, v. 56, n.

2, p. 225-236, 1997a.

MARTINS, M. B. G. ; CASTRO, P. R. C. Biorreguladores na morfologia e na produtividade de frutos de tomatoiro cultivar Angela Gigante. **Bragantia**, Campinas, v. 56, n. 2, p. 237-248, 1997b.

MEDEIROS, M.A.; VILLAS BOAS, G.L.; VILELA, N.J.; CARRIJO, O.A. Estudo preliminar do controle biologico da traga-do-tomateiro com o parasitoide Trichogramma pretiosum em ambientes protegidos. **Horticultura Brasileira**. v.27, n.1, p.83-85, 2009.

MINAMI, K.; HAAG, H.P. **O Tomateiro.** 2.ed. Campinas: Cargill Foundation, 397p. 1989.

MOMOL MT, OLSON SM, FUNDERBURK JE, STAVISKY J, AND MAROIS JJ. **Integrated management of tomato spotted wilt on field-grown tomatoes**. Plant Disease 88:882-890. 2004.

MOUND, L..A. The Thysanoptera Vector Species Of Tospoviruses. **Acta Hort**. (ISHS) 431:298309. 1996. Available at http://www.actahort.org/books/431/431_25.htm. Accessed 23 June 2013.

NAIKA, S.; JEUDE, J. L. J.; GOFFAU, M.; HILMI, M.; DAM, B. **The tomato crop: production, processing and marketing.** [S.l.: s.n., 2009?].

PAPADOPOULOS, A.P., PARARAJASINGHAM, S. The influence of plant spacing on light interception and use in greenhouse tomato *(Lycopersicon esculentum* Mill.): a review. **Scientia Horticulturae**, Amsterdam, v.69, p.1-29, 1997.

PAZINATO, B.C.; GALHARDO, R.C. **Processamento artesanal do**

tomate. Campinas: Coordenadoria de Assistencia Tdcnica Integral, 30 p. 1997.

REITZ, S. R. Comparative bionomics of *Frankliniella occidentals* and *Frankliniella tritici*. **Florida Entomologist** v. 91, p. 474-476, 2008.

RILEY, D.G. AND PAPPU, H. R. Evaluation of tactics for management of thrips-vectored tomato spotted wilt tospovirus in tomato**. Plants Diseases**. v. 84, p. 847-852, 2000.

RILEY, D.G. AND PAPPU, H. R. Tactics for management of thrips (Thysanoptera, Thripidae) and Tomato spotted wilt tospovirus in tomato. J. **Economy Entomology**. v. 97, p. 1648-1658, 2004.

RILEY, D.G., R. MCPHERSON AND L. WELLS. **"Thrips vectors of TSWV,** pp 13- 16.In: 2nd revision of tosporviruses in Solanaceae and other crops in the Coastal Plain of Georgia. University of

Georgia CAES Research Report Number 704 (In press). 2009a.

RILEY, D.G., R. MCPHERSON, L. WELLS AND S. BROWN. **"Management of thrips vectors of TSWV**, pp 24-26. In: 2nd Revision of Tospoviruses in Solanaceae and other crops in the Coastal Plain of Georgia, University of Georgia CAES Research Report Number 704 (In press). 2009b.

RILEY, DAVID G. "A reduced risk system for managing thrips and TSWV in tomato and pepper", pg 25. In: W.T. Kelly (Ed) **Proceedings of the SE Regional Vegetable Conference**, 54 pp. 2008.

SCOTT, S.J., MCLEOD, P.J., MONTGOMERY, F.W., HANDLER, C.A. Influence of reflective mulch on incidence of thrips (Thysanoptera: Thripidae:Phlaeothripidae) in staked tomatoes. **Journal Entomology**. Sci. v. 24, p. 422-427, 1989.

SHERWOOD JL; GERMAN TL; WHITFIELD AE; MOYER JW; ULLMAN DE. Tomato spotted wilt. In: ***ENCYCLOPEDIA of plant pathology***. New York: John Wiley. p.1030-1031. 2001a.

SHERWOOD, J. L., T. L. GERMAN, A. E. WHITFIELD, J. W. MOYER, AND D. E. ULLMAN. **Tospoviruses**, pp. 1034-1040. In Encyclopedia of Plant Pathology. John Wiley & Sons, Inc. New York. 2001b.

SILVA, F. A. S.; AZEVEDO, C. A. V. Versao do programa computacional Assistat para o sistema operacional Windows. Revista Brasileira de Produtos Agroindustriais, v. 4, n. 1, p. 71-78, 2002.

SILVA, J.B.C.; GIORDANO, L.B. **Tomate para processamento industrial.** Brasilia: Embrapa Comunicagao para Transferencia de Tecnologia - Embrapa Hortaligas, 168 p. 2000.

SONNENBERG, P. E.; SILVA, N. F. **A cultura do tomate**. Olericultura especial: as culturas de: alface, cenoura, batata, tomate, cebola e alho. 8. ed. Goiania: Universidade Federal de Goias, 2004. p. 49-90.

SRIVISTAVA, M., L. BOSCO, J. FUNDERBURK, S. OLSON, AND A. WEISS... Spinetoram is compatible with the key natural enemy of *Frankliniella* species thrips in pepper. **Plant Health Progress** doi:10.1094/PHP-2008-0118-02-RS. 2008

STAVISKY, J., FUNDERBURK, J.E., BRODBECK, B.V., OLSON, S.M., ANDERSEN, P.C. Population dynamics of Frankliniella spp. and tomato spotted wilt incidence as influenced by cultural management in tomato. **Journal Economy Entomologist**. v. 95, p. 1216-1221, 2002.

STRECK, A. A.. **Occurrence and action of harmful insects in different vegetable cultivars in Cachoeira do Sul, RS, Brazil**. Fac. Filos. Cienc. Letras, Cachoeira do Sul, 67p. 1994

TARARA, J.M. Microclimate modifications of plastic mulch. **HortScience** v.

35, p.169-180, 2000.

WEISS, A., J. E. DRIPPS, AND J. E. FUNDERBURK.. Assessment of implementation and sustainability of integrated pest management programs. **Florida Entomologist.** v. 92, p. 2428, 2009

ZORZOLI, R.; PRATTA, G.R.; PICARDI, L.A. Variabilidad gendtica para la vida postcosecha y el peso de los frutos en tomate para familias F3 de un hlbrido interespeclfico. **Pesquisa Agropecuaria Brasileira,** Brasilia, v. 35, n. 12, p. 2423-2427, Dec. 2000.

ZUCCHI, R. A.; SILVEIRA NETO.; NAKANO, O. **Guia de identifica^ao de pragas agricolas**.

Piracicaba: FEALQ, 1993. p. 139.

CHAPTER 7

APPENDICES

APPENDIX 1 - Graph of the electromagnetic spectrum in the range 350nm to 400nm, of the polyethylene plastic in white colour. Prepared by the ViewSpec Pro.INK software.

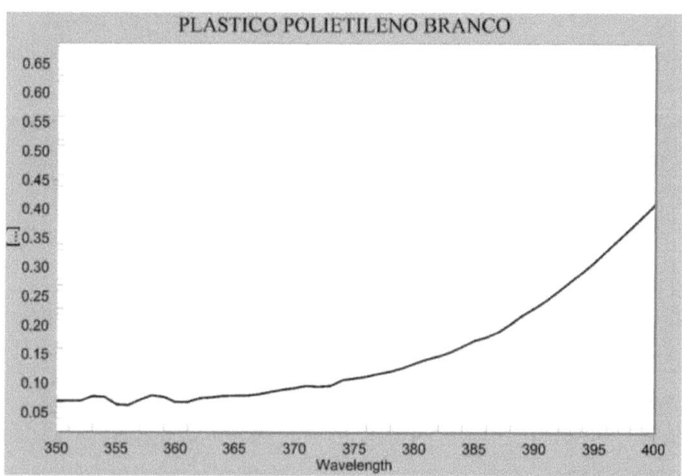

APPENDIX 2 - Graph of the electromagnetic spectrum in the range 350nm to 400nm, of the polyethylene plastic in yellow colour. Prepared by the ViewSpec Pro.INK software.

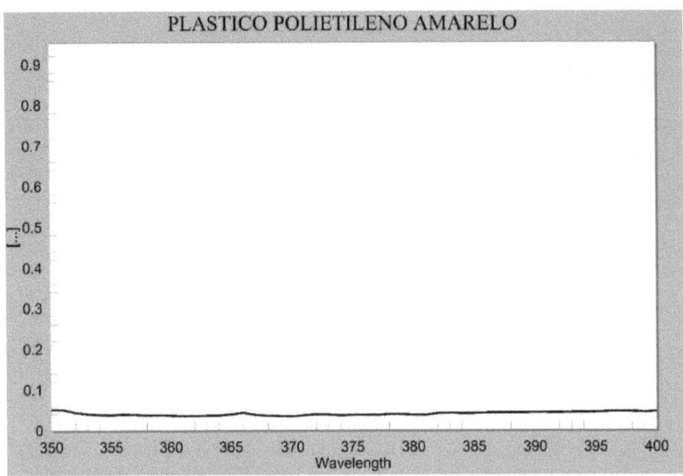

APPENDIX 3 - Graph of the electromagnetic spectrum in the range 350nm to 400nm, of the polyethylene plastic in grey colour. Prepared by the ViewSpec Pro.INK software.

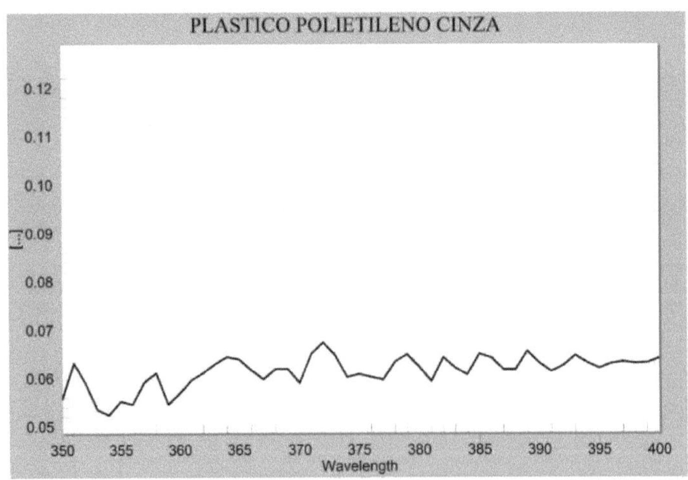

APPENDIX 4 - Graph of the electromagnetic spectrum in the range 350nm to 400nm, of the polyethylene plastic in silver colour. Prepared by the ViewSpec Pro.INK software.

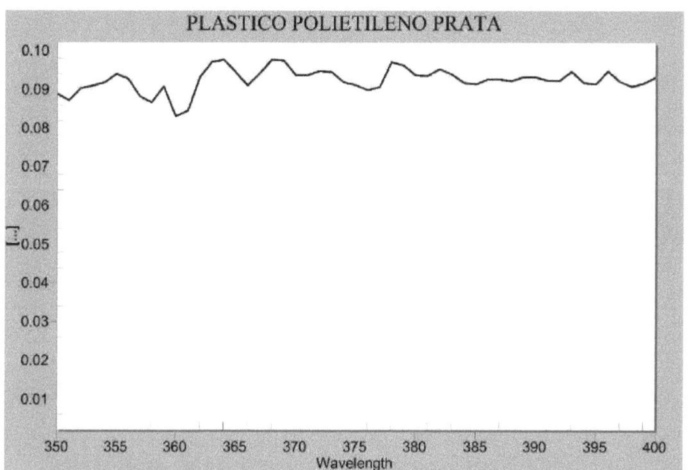

APPENDIX 5 - Graph of the electromagnetic spectrum in the range 350nm to 400nm, of the polyethylene plastic in red colour. Prepared by the View Spec Pro.INK software.

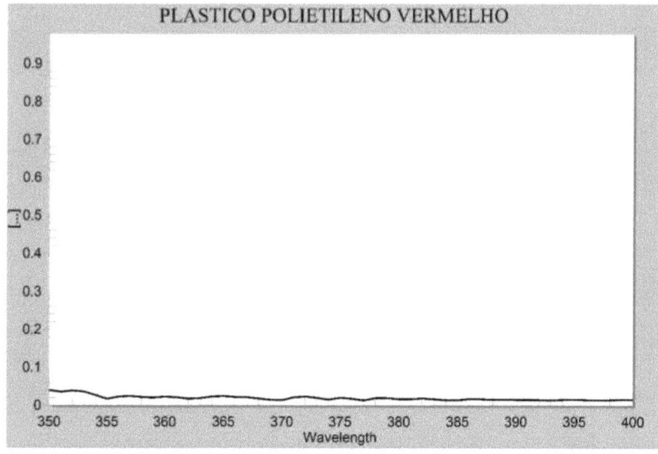

APPENDIX 6 - Graph of the electromagnetic spectrum in the range 350nm to 400nm, of the polyethylene plastic in black colour. Prepared by the ViewSpec Pro.INK software.

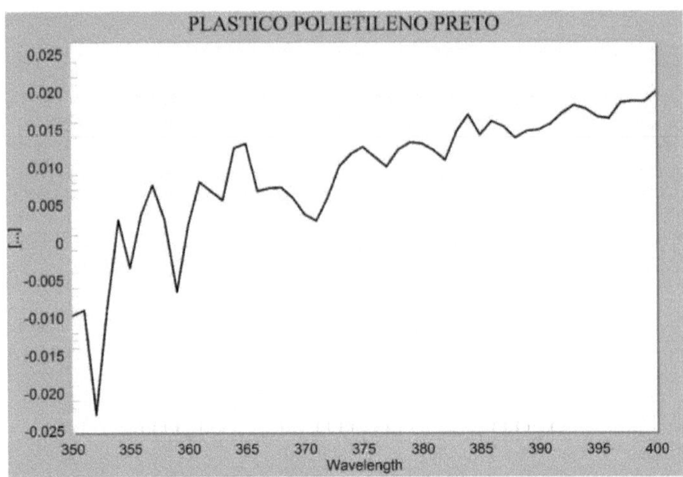

APPENDIX 7 - Graph of the electromagnetic spectrum in the range 350nm to 400nm, of rice straw. Prepared by the ViewSpec Pro.INK software.

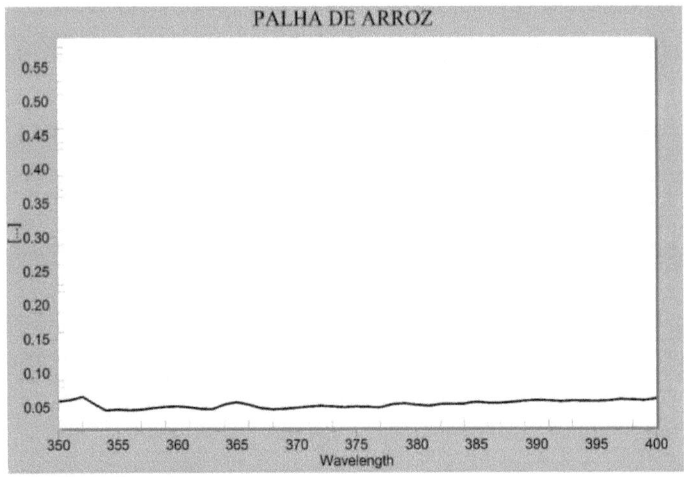

APPENDIX 8 - Graph of the electromagnetic spectrum in the range 350nm to 400nm, of the wet soil. Prepared by the ViewSpec Pro.INK software.

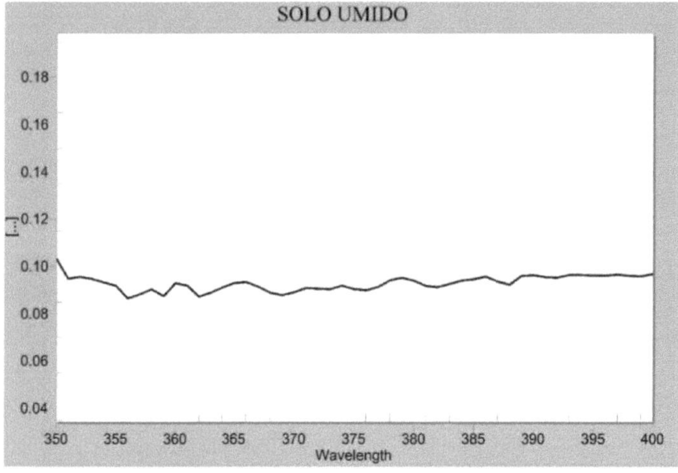

Printed by Books on Demand GmbH, Norderstedt / Germany